10-14-75

WINDMILLS

KVARMEN, SWEDEN

WINDMILLS
by FRANK BRANGWYN·RA
AND HAYTER PRESTON

London: JOHN LANE THE BODLEY HEAD LIMITED
New York: DODD, MEAD AND COMPANY

Republished by Gale Research Company, Book Tower, Detroit, 1975

First Published in 1923

Library of Congress Cataloging in Publication Data

Preston, Hayter, 1891-
  Windmills.

    Reprint of the 1923 ed.
    1. Windmills. I. Brangwyn, Sir Frank, 1867-1956, illus. II. Title.
TJ823.P7   1975   621.4'5   70-178621
ISBN 0-8103-4077-1

CONTENTS

	Page
Introduction	11
Barking	33
Ramsgate	39
Winchelsea	47
Littlehampton	53
Campo de Crijitano	59
Keston	65
Bruges	71
Weerdt, Holland	77
Herne	85
Whitstable	89
Dunkirk	95
Aldeburgh	99
Sark	105
Crowborough	109
Montreuil-sur-Mer	115
Cologne	121

LIST OF ILLUSTRATIONS

Kvarmen, Sweden	Frontispiece
	Facing Page
A Sussex Mill	12
Molen de Koker, Zaandam	16
Mills, Öland	22
Montreuil-sur-Mer	28
Barking Mill	34
Last of Ramsgate Mill	40
Winchelsea Mill	48
The Mills of Crijitano	60
Furnes, Belgium	72
A Pumping Mill, Holland	76
Weerdt, Holland	78
A Kentish Mill	86
St. Bennett's Abbey Mill	98
Crowborough	110
Cologne	122

MAN, in his effort to harness the forces of nature, has produced two beautiful wind-drinking contrivances—the windmill and the sailing ship. There is a beauty, individual and uncounterfeitable, in every sailing ship and windmill, a charm direct and personal as the charm of a friend.

In spite of their solid bodies and mundane purposes, there is about both of them an indefinable spirit, something elemental, fluid, other-worldly. A sailing ship traversing the ocean is to me as wonderful and mysterious as a meteor crossing the heavens, and windmill sails revolving against the blue and green of quiet lands arouse in me feelings as deep and mystical as those with which I regard the remote and whirling stars.

The sailing ship and the windmill are both essentially romantic creations. The one defying and thwarting the ungovernable fury of the sea; the other stemming the tide of the mighty wind with its frail fingers.

INTRODUCTION

A SUSSEX MILL

WITH regard to the early history of Windmills, I cannot do better than quote from a work by Professor John Beckmann. Dealing with "Corn Mills," he says: "The intrusting of that violent element water to support and drive mills constructed with great art, displayed no little share of boldness; but it was still more adventurous to employ the no less violent but much more untractable and always changeable wind for the same purpose. Though the strength and direction of the wind cannot be anyway altered, it has, however, been found possible to devise means by which a building can be moved in such a manner that it shall be exposed to neither more nor less wind than is necessary, let it come from what quarter it may.

"It is very improbable—or, much rather, false—that the Romans had windmills; though Pomponius Salinus affirmed so, but without any proof(1). Vitruvius(2), where he speaks of all forces, mentions also the wind; but he does not say a word of windmills. Nor are they noticed by Seneca(3) or Chrysostom(4) who have both spoken of the advantages of the wind. I consider as false also the account given by an old Bohemian annalist(5), who says that before the years 718

(1) See Pomponius Sabinus, *ut supra*.   (2) Lib. ix. c. 9, lib. xc. 1, 13.
(3) Natur. Quaest., lib. v. c. 18.   (4) Chrysost. in Psalm cxxxiv. p. 362.

(5) "At the same period (718), one named Halek, son of Uladi the Weak, built close to the city an ingenious mill which was driven by water. It was visited by many Bohemians, in whom it excited much wonder, and who, taking it as a model, built others of the like kind here and there on the rivers; for before that time all the Bohemians' mills were windmills erected on mountains."

there were none but windmills in Bohemia, and that watermills were then introduced for the first time. I am of opinion that the author meant to have written hand and cattle instead of windmills.

"It has been often asserted that these mills were first invented in the East, and introduced into Europe by the Crusaders; but this also is improbable, for mills of this kind are not at all, or very seldom, found in the East. There are none of them in Persia, Palestine, or Arabia; and even watermills are there uncommon, and constructed on a small scale.

"Besides, we find windmills before the Crusades, or at least at the time when they were undertaken. It is probable that these buildings may have been made known to a great part of Europe, and particularly in France and England(1), by those who returned from these expeditions; but it does not thence follow that they were invented in the East(2).

(1) See De la Mare, Traité de la Police, etc., *ut supra*; Description du Duché de Burgoyne, Dijon, 1775, 8vo., 1. p. 163; Dictionnaire des Origines, par D'Oriquy, v. p. 184. The last work has an attractive title; but it is the worst of its kind, written without correctness or judgment, and without giving authorities.

(2) "There are no windmills at Ispaham nor in any part of Persia. The mills are all driven by water, by hand, or by cattle." (*Voyages de Chardin*, Rouen, 1723, 8vo., viii. p. 221). "The Arabs have no windmills; these are used in the East, only in places where no streams are to be found. And in most parts the people make use of handmills. Those which I saw on Mount Lebanon and Mount Carmel had a great resemblance to those which are found in many parts of Italy. They are exceedingly simple, and cost very little. The millstone and the wheel are fastened to the same axis. The wheel, if it can be so called, consists of eight hollow boards, shaped like a shovel, placed across the axis. When the water falls with violence upon these boards, it turns them round, and puts in motion the millstone, over which the corn is poured." (D'Arvieux:

"The Crusaders perhaps saw such mills in the course of their travels through Europe; very probably in Germany, which is the original country of most large machines. In like manner, the knowledge of several useful things has been introduced into Germany by soldiers who have returned from different wars; as the English and French, after their return from the last war, made known in their respective countries many of our useful implements of husbandry, such as our straw-chopper, scythe, etc.

The modest Prussian!

"Mabillon mentions a diploma of the year 1105, in which a convent in France is allowed to erect water and wind mills, *molendina ad ventum*(1). In the year 1143, there was in Northamptonshire an Abbey, situated in a wood, which in the course of one hundred and eighty years was entirely destroyed. One cause of the destruction was said to be, that in the whole neighbourhood there was no house, wind or watermill built, for which timber was not taken from this wood(2).

"In the twelfth century, when the mills began to be more

Merkwürdige Nachrichten von seinen Reisen, Part III., Copenhagen and Leipsic, 1754, 8vo., p. 201). "I did not see either wind or water mill in all Arabia. I, however, found an oil-press at Téhama, which was driven by oxen, and thence suppose that Arabs have corn-mills of the like kind." (*Niebuhr Beschreibung von Arabien*, p. 217).

(1) "Iisdem etiam facultatem consexit constituendi domos, staqna, molendina ad aquam et ventum, in episcopatu Ebroicensi, Constantiensi, et Bajocensi, ad angendos monasterü proventus."

(2) Prœterea non fuit in patria, aula, camera, ovreum, molendinum venticium sive aquaticum alicujus valoris plantata sine adminiculo aliquo boscorum Sanctæ Mariæ de Pipewalla (so the wood was called) quot virgæ molendinorum venticiorum dabauntur in temporibus di versorum

common, a dispute arose whether the tithes of them belonged to the clergy; and Pope Celestine III. determined the question in favour of the Church(1). In the year 1332, one Bartolommeo Verde proposed to the Venetians to build a windmill. When his plan had been examined, a piece of ground was assigned to him, which he was to retain in case his undertaking should succeed within a time specified(2). In the year 1393 the city of Spires caused a windmill to be erected, and sent to the Netherlands for a person acquainted with the method of grinding by it(3).

"A windmill was also constructed at Frankfort in 1442, but I do not know whether there had not been some there before(4).

"To turn the mill to the wind, two methods have been invented. The whole building is constructed in such a manner as to turn on a post below, or the roof alone, together with the axle-tree; and the wings are movable. Mills of the former kind are called German Mills; those of the latter,

abbatum nemo novit, nisi Deus. Caussa tertia destruction is boscorum fruit in constructione et emendatione domorum infra abbathiam et extra utpote grangus, orrcis bercarüs molendinis aquaticis et venticüs per vices. (The letter of donation, which appears also to be twelfth century, may be found in the same collection, vol. ii. p. 459. In it occurs the expression, *molendinum ventritricum*. In a character, also, in vol. iii. p. 107, we read of *molendinum ventorium*. (*Monasticon Anglicanum sive Pandictæ Cœnobiorum*, edit. sec. London, 1682, fol., i. p. 816).

(1) De reditibus molendini ad ventum solvendæ sunt decimæ, Secretal, Greq., lib. iii. tit. 30, c. 23.

(2) Gir. Zanetti, Dell' friqine di alcune arte appresso di Veneziani, Veney., 1758, 4to., p. 74; Pro faciendo unum molandinum a vento; Le Bret, Geschichte von Venedig, II. i. p. 233.

(3) Lehmann's Ckronica de Stadt Spezger. Frankf. 1662, 4to., p. 847: "Sent to the Netherlands for a miller who could grind with the windmill."

(4) Lemner, Frankf. Chronik, ii. p. 22

MOLEN DE KOKER, ZAANDAM

Dutch. They are both moved round, either by a wheel and a pinion within, or by a long lever without(1). I am inclined to believe that the German windmills are older than the Dutch; for the earliest descriptions which I can remember, speak only of the former. Cardan(2), in whose times windmills were very common, both in France and Italy, makes, however, no mention of the latter; and the Dutch themselves affirm, that the mode of building with a movable roof was first found out by a Fleming in the middle of the sixteenth century(3).

"Those mills by which, in Holland, the water is drawn up and thrown off from the land, one of which was built at Alkmaar in 1408, another at Schoonhoven in 1450, and a third at Enkluysen in 1452, were at first driven by horses, and afterwards by wind. But as these mills were immovable,

(1) Description and figures of both kinds may be found in Leupold's Theatrum Machinarum Generale, Leipsic, 1724, fol., p. 101, tab. 41-43.

(2) "Nor can I pass over in silence what is so wonderful, that, I could neither believe nor relate it, though commonly talked of, without incurring the imputation of credulity. But a thirst for science overcomes bashfulness. In many parts of Italy, therefore, and here and there in France, there are mills which are turned round by the wind." (*De Rerum Varietate*, lib. i. cap. 10, in the edition of all his works, Lugduni, 1663, fol., vol. iii. p. 26).

(3) This account I found in De Koophandel van Amsterdam, door le Lang, Amsterdam, 1727, 2 vols., 8vo., 11, p. 584: "*De beweegelyke kap, om de moolens op all windens te zettens, is erst in't midden van de xvide eeuw door een Vlaaming uytgevonden.*" ("The movable top for turning the mill round to every wind was first found in the middle of the sixteenth century by a Fleming"). We read there that this is remarked by John Adrian Leegwater; but of this man I know nothing more than what is related of him in the above work, that he was celebrated on account of various inventions, and died in 1650, in the seventy-fifth year of his age. See also Beschryving der Stadt Delft door verscheide Liefhebbers en Kenners der Nederlandsche oudhedin. Te Delft, 1729, fol., p. 623.

and could work only when the wind was in one quarter, they were afterwards placed, not on the ground, but on a float which could be moved round in such a manner that the mill should catch every wind(1). This method gave rise, perhaps, to the invention of movable mills."

In Jocelin's Chronical there is a mention of a mill as a cause of dispute. Generally mills were the property of the baron, and the peasants who used the mills for grinding their corn had to pay. One Dean Herbert built a mill without the consent of the Abbot, who stated: "I tell thee, it will not be without damage to my mills; for the townsfolk will go to thy mill, and grind their corn at their own good pleasure; nor can I hinder them, since they are free men. I will allow no new mills on such principle." The Abbot ordered the mill to be taken down, but when his men arrived they found that the Dean had already dismantled the mill, so that he might not lose the timber. This certainly suggests that it was a windmill and not a watermill.

Disputes arising out of windmills appear to have been frequent, and an episode is given by Polckmann in connection with the use of windmills, which may be of interest to economists who believe that natural forces should be common property:

"The avarice of landholders, favoured by the meanness

(1) De molens hadden doen (toen) vaste kappen zoo datze maar met eene wind malen konde, waar om men op zekere plaats, om dit ongeval voor te kommen, een molen op een groot vlot weder zette dat men dan maar din wind draide." See the History of the City of Delft, above quoted.

and injustice of governments, and by the weakness of the people, extended the regality not only over all streams, but over the air and the windmills. The oldest example of this with which I am acquainted is related by Jargow(1).

"In the end of the fourteenth century, the monks of the celebrated but long since destroyed monastery of Augustines at Windsheim, in the province of Overyssel, were desirous of erecting a windmill not far from Zwoll; but a neighbouring lord endeavoured to prevent them, declaring that the wind in the district belonged to him.

"The monks, unwilling to give up their point, had to recourse to the Bishop of Utrecht, under whose jurisdiction the province had continued since the tenth century. The Bishop, highly incensed against the pretender who wished to usurp his authority, affirmed that the wind of the whole province belonged to him only, and in 1391 gave the convent express permission to build a windmill wherever they thought proper(2).

"In like manner, the city of Haerlem obtained leave from

(1) Jargow, *Einleitung in die Lehre von den Regalien,* Rostock, 1757, 4to., p. 494.

(2) "As our monastery had not a mill to grind corn, they resolved to build a new one. When the lord of Woerst heard this, he did every thing in his power to prevent it, saying that the wind in Zealand belonged to him, and no one ought to build a mill there without his consent. The matter was therefore referred to the Bishop of Utrecht, who, as soon as the affair was made known to him, replied in a violent passion that no one had power over the wind within his diocese but himself and the Church at Utrecht; and he immediately granted full power, by letters-patent, dated 1391, to the convent at Windsheim, to build for themselves and their successor a good windmill in any place which they might find convenient." (*Chronicon Canonicorum regularium ordinis Augustini, capituli Windesemensis, auctore Joh. Buschio,* Antveripæ, 1621, 8vo., p. 73.)

Albert, count palatine of the Rhine, to build a windmill, in the year 1394.

"Another restraint to which men in power subjected the weak, in regard to mills, was, that vassals were obliged to grind their corn at their lord's mill, for which they paid a certain value in kind. The oldest account of such ban-mills, *molendina bannaria,* occurs in the eleventh century. Fulbest, Bishop of Chartres and Chancellor of France, in a letter to Richard, Duke of Normandy, complains that attempts began to be made to compel the inhabitants of a part of that province to grind their corn at a mill situated at a distance of five leagues(1). In the chronical of the Benedictine monk Hugo de Flavigny, who lived in the eleventh and twelfth century (sic), we find mention of *molendina guatuor cum banno ipsius villœ*(2). More examples of this servitude, *secta ad molendinum,* in the twelfth and thirteenth centuries, may be seen in Du Fresne, under the words *molendinum bannale.*

"It is not difficult to account for the origin of these ban-mills. When the people were once subjected to the yoke of slavery, they were obliged to submit to more and severer servitudes, which, as monuments of feudal tyranny, have continued even to more enlightened times. De la Mare(3) gives

---

(1) "Albetus notum facimus quod donavimus donamusque civitati nostræ Harlemianæ ventum molarium a parte australi civitatis nostræ praiscriptæ hemistadium versus inter Pacis forsam et sparnam." (*Theo. Schrevelü: Harlemum Lugduni, Batavorum,* 1647, 4to., p. 181).

(2) This letter of Fulbert may be found in Maxima Bibliotheca Veterum Patrum. Lugduni 1677, fol., tom. xviii. p. 9.

(3) In Labbei Biblioth, Manuser. i. p. 132.

an instance where a lord, in affranchising his subjects, required of them, in remembrance of their former subjection, as that he might draw as much from them in future as possible, that they should agree to pay a certain duty, and to send their corn to be ground at his mill, their bread to be baked in his oven, and their grapes to be pressed at his winepress. But the origin of these servitudes might perhaps be accounted for on juster grounds. The building of mills was at all times expensive, and undertaken only by the rich, who, to indemnify themselves for the money expended in order to benefit the public, stipulated that the people in the neighbourhood should grind their corn at no other mills than those erected by them."

In spite of Backmann, I am inclined to believe that Dutch mills are the older of the two. Apart from corn-grinding, windmills have been for centuries used in Holland for draining purposes. There are upwards of 2,000 windmills helping to drain some 2,000,000 acres. The only means the early Dutch had of keeping dry the land reclaimed from the sea was the windmill; but now, of course, the steam-pump has taken the place of the mill; for though the wind may be a power which is at man's service free of charge, it is a whimsical friend that often refuses its service when it is most needed.

Windmills are in use in all parts of the world in various capacities. In Barbados there are a few mills of the old Dutch or Tower type, and they are worked for crushing sugar-cane. In Algeria there are a few windmills used for pumping water; in Cape Colony they are used most for watering stock; and the two mills on St. Helena do most of the threshing of the island. There are comparatively few windmills in China; it is said that native superstitution is against their use. I have however, seen a photograph of a Chinese windmill—a weird, bamboo skeleton, circular in shape, with strips of cloth to catch the wind. It was apparently used for pumping water.

The older European windmills can be divided into two general types—the horizontal mill and the vertical mill. In the former type the sails are so placed as to turn by the force of the wind in a horizontal plane, and about an axis nearly vertical. In the latter type the sails turn in an almost vertical plane, that is, about an axis nearly horizontal.

The horizontal mill has many disadvantages, and, consequently, their use has been limited. They were generally used in positions in which the height of vertical sails were an obstacle, but such an objection was exceedingly rare. The general construction of the horizontal mill is thus described by Wolff: Six or more sails, consisting of plane boards, are set upright upon horizontal arms which rest upon a tower, and which are attached to a vertical shaft passing through the centre of the tower. The sails, which are fixed in position, are set obliquely to the direction in which the wind will strike them. Outside of the whole is placed a screen or cylindrical arrangement of board intended to revolve, these

MILLS, ÖLAND

boards being set obliquely, and in planes lying in opposite course to those of the sail. As a result, from whatever direction the wind may blow against the tower, it is always admitted by the outer boards to act on the sails most freely in that half of the side it strikes on, from which the sails are turning away; and it is partly, though by no means entirely, broken from the sails which, in the other quadrant of the side, are approaching the middle line. Fairbairn gives the following account of a horizontal windmill at Eupatoria, in the Crimea as it appeared during the period of the Crimean War. This description will well answer for the whole type:

"Around the town of Eupatoria, in the Crimea, there appeared to be nearly two hundred windmills, chiefly employed in grinding corn; and all which were in a workable state were of the vertical construction, and only one horizontal mill, which seemed to have been out of use for at least a quarter of a century. The tower of this mill was built of brickwork, about twenty feet diameter at the base, and about seventeen feet at the top, and twenty feet high. The revolving wings, which consisted of six sets of arms, appeared to be about twenty feet diameter and about six feet broad, fitted with vertical shutters which were movable on pivots passing through the arms, the shutters being about twelve inches wide by five or six feet high; and the pivots were fixed at about one-third of the breadth from the edge of the shutters, in order that the wind might open and shut them at the proper time during the revolution of the wings. About one-third of the circumference of the wings was surrounded by a segmental screen, to shelter the arms and shutters, while moving up against the wind; and the screen seemed to have been hauled round with ropes, in order to suit the direction of the wind."

The objections to the employment of the horizontal windmill, which virtually debarred, and still debars it from use in competition with the vertical mill are: first, that only one or two sails can be effectually acted upon at the same moment; and secondly, that the sails move in a medium of nearly the same density as that by which they are impelled, and that therefore great resistance is offered to those sails which approach the middle. Smeaton, in his *Philosophical Transactions*, 1755 to 1763, says: "Little more than one sail can be acting at once, whereas in the common windmill all the four act together; and, therefore, supposing each vane of a horizontal windmill has the same dimensions as each vane of the vertical, it is manifest that the power of a vertical mill with four sails will be four times greater than the power of the horizontal one, let its number of vanes be what it will. This disadvantage arises from the nature of things; but, if we consider the further disadvantage that arises from the difficulty of getting the sails back against the wind, etc., we need not wonder if this kind of mill is in reality found to have not above one-eighth to one-tenth of the power of the common sort, as has appeared in some attempts of this kind."

While it is true, that, with a like area of sails, the power of the horizontal is always much less than that of the vertical mill, Smeaton's estimate of one to eight or one to ten is too unfavourable, inasmuch as he overlooked, as Sir David Brewster first showed, the loss in vertical mills of one component part of the wind's pressure. The ratio of one to four, given by Sir David Brewster, is, however, about the correct figure, and presents a sufficient explanation of the limited use to which horizontal windmills have been put in the past, and a sufficient cause why they should not be employed at the present time, if the question of economy of motive power at all enters the problem as a leading consideration.

Of vertical mills the same authority says: "In vertical mills . . . the tower or building which supported the windmill proper was either of wood or stone: if of stone, the tower was commonly in the form of a frustum or cone. The principal parts of the mills proper are:—

(1). An axle or shaft, either of wood or iron, in the top of the building, inclined to the horizontal at an angle of from ten to fifteen degrees, as observation has shown that the impulse of the wind is usually exerted in lines descending at such angles.

(2). The sails attached to near the outer extremity of the shaft, and turning in nearly a vertical plane. The planes of these sails are placed obliquely to the plane of the revolution; so that, when the wind blows in the direction of the axle, it impinges upon their surface obliquely, and thus the effort of the sail to recede from the wind causes it to turn upon its axle. These sails consist of wooden frames (arms and crossbars), with canvas covering the lattice or frame work. If four in number, as is the rule, though five and six have been employed, the sails are fixed in position at right angles to each other. They are usually constructed from thirty to forty feet in length, though fifty feet has often been exceeded.

(3). A large toothed wheel upon the horizontal axle, the teeth of which engage with those of a pinion upon

(4). A vertical shaft from which motion is imparted to the machinery.

It will be understood that the horizontal shaft is supported at its inner end near the centre of the base of the dome or cone surmounting the mill, while its opposite extremity passes through a perforation in one side of the dome, where it has its main support, and projects far enough to receive the ends of the long timbers or arms of the sail. The pivot at the lower or inner end of the shaft takes up but a small part of the weight and counter-pressure.

The axle is constructed of some hard wood, like oak, or of wrought-iron with cast-iron flanges of large diameter keyed on the front, which are furnished with recesses for receiving and holding the arms of the sails. The latter must be proclaimed as the better practice; since, the diameter of the neck of the wooden shaft being from one and a half to two feet, an iron one substituted in its place need not be more than six to nine inches, and thus the loss by friction is materially decreased.

The sails are made plain, concave, or warped. The latter, the most effective, have been in greatest use; and the angles employed in the Dutch type of mill have, on the whole, approached very closely to those which theoretical analysis proves to be most serviceable. Where plane sails have been used, the bars have all had the same angle of inclination, ranging between twelve and eighteen degrees to the plane of revolution. It is not necessary to say more in a general way about the sails than that they are either of rectangular or (more usually) of trapezoidal form, increasing in width as they approach the outer extremity of arm; that the innermost cross-bar is placed at about one-sixth to one-seventh of the length of the arm from the middle of the shaft; and that its length is about equal to this distance. So the canvas lattice-work covers only five-sixths or six-sevenths of the outer portion of the sails. In a sail about thirty feet long, the arms near the shaft are about one foot thick and nine inches wide, and at the outer end about six inches thick and four and a half inches wide.

As the direction of the wind is changing perpetually, some contrivance is necessary for bringing the shaft into the direction of the wind, so that the sails will be acted upon most effectively. According as this revolution is effected, European vertical windmills have been divided into two general types:—

1. The *Post or German Mill*, in which the whole building which sustains the wind sails, shaft, and the machinery is supported upon a vertical post or column, upon which it revolves at will when actuated by a lever.

2. *The Tower or Dutch Mill*, in which only the head, cap, or dome of the building, with the shaft which it contains, revolves.

Post or German Mills. It will be readily understood that not only are these mills necessarily limited in their size, but that the manual labour their turning to the wind implies, led to their effectual abandonment when the tower mills had been made automatic in their regulation.

The following description of a *Post Mill* by Peter Barlow, F.R.S., gives a good idea of the construction:

"One end of the wind shaft has a bearing on the beam of the framing of the mill, and the other is supported in a similar way by a beam; the part of the shaft outside the mill is larger, and made square, and has two square holes or mortises through it, into which the whips or arms of the sails are fitted and made fast by wedges. The wheel which is termed the brake-wheel, is attached to the wind-shaft; it has a rim of wood on its circumference, termed the brake, one end of which is attached to a fixed part of the mill, and the other, by means of an iron rod, to a lever; so that, by pressing down on the end of the lever, the brake is made to bind upon the circumference of the wheel, and thereby produce such a resistance that the mill may be at any time stopped.

"The lower floor of the mill is made to receive the post upon which the mill is turned round to face the wind. This post is a very strong tree, which is held perpendicularly by fixing it upon the middle of two long timbers, which form a large cross upon the ground, and which constitute the base of the whole mill. The post is secured in its vertical position by four oblique braces which extend from the ground across to the middle of it; leaving ten or twelve feet of the upper

MONTREUIL-SUR-MER

part, which is made round, clear from the obstruction of the braces. This round part of the post rises up through the middle of the lower chamber, in the floor of which a circular collar is formed to the exact diameter of the post. At the other end of the post is a pivot or gudgeon, which enters into a socket fixed to one of the strongest beams in the middle of the upper floor; this beam must necessarily be very strong, as it has to sustain the whole weight of the erection. In this way the mill is made to turn freely upon the pivot, while the collet in the lower floor serves to keep it steady and in a vertical position. There is a ladder for the purpose of ascending to the mill: it is united by joints to the back of the framing, and has a rope fastened to the lower end, which passes in an inclined direction into the mill, so that, by a lever or pulley, it can be raised at pleasure clear of the ground. The ladder thus raised serves as a lever for turning the mill round, which is usually done by manual labour: sometimes, however, more force is necessary, and a small capstan is provided, to draw a rope attached to the end of the ladder. This capstan is movable, and can be fastened at pleasure to any of the posts which are fixed in the ground for the purpose. When the mill is by these means placed in the desired direction, the ladder is let down to the ground; and, its position being the opposite side to that of the sails, it serves not only for ascent, and to keep the mill steady in position, but acts as a stay to resist the tendency of the wind to overturn it—an occurrence which sometimes happens in mills of this description."

WINDMILLS

THE MILL, BARKING

BARKING MILL

I STARTED out one Sunday morning for a ramble around the eastern side of London. A blue sky touched the grey squalor of Whitechapel and Poplar with prismatic lights; the malodorous air was slashed with shafts of sunlight; and the Sunday crowd stayed in one's mind with an intense vividness. Somebody had told me of a windmill behind this "desert of brick and stone." And it was this windmill that I sought.

I found it after an hour's journey, and the sight of it was a shock. It stands out black and stark among the half-ruined greenness of the Barking Road, a monument of defiant, desperate impotence—the grim anatomy of a Titan that in the fullness of its strength ground one hundred sacks of wheat weekly, besides grinding most of the cattle corn for neighbouring farmers; now reduced by the elements and the changing fashions of man to cutting occasional sacks of chaff with the aid of an electric dynamo!

A low dwelling slouched like a drunken man at the foot of the Mill, and through the window I saw a white head nodding to a black shadow at the end of the room.

I knocked and was welcomed in with a simple courtesy that belongs, I fear, to another age. I wanted to talk about the Mill. Certainly. Would I come in? I went in.

An old man was sitting under the window, a notable and tragic figure. A shock of white hair covered the noble, toil-lined face, and his twisted limbs and the two sticks between his knees told their own tale. He was paralysed. The huge wingless Mill threw dark stains on the walls and ceiling; and the helpless old man cast a shadow on my spirit.

He was glad to talk to somebody about the Mill. Few folks had interest in mills nowadays; they had gone, he said, like these. . . . And he pointed to his legs.

The old woman in a poke bonnet who had shown me in smiled half-pityingly with the clearest eyes in the world. She was his wife.

It would take a long time to tell me about the Mill, she said: and I might like "a bite." Did I like mutton and peas and potatoes? . . . And a little mint sauce? . . . I felt like a child. And when my eyes fell upon the old man again, I understood. This kindness was her business in life.

He told me many things of the remote and distant past; quiet, half-plaintive memories of the old days. He had viewed life from a windmill and seemed to have caught something of its detachment from the earth. And the Mill has left its marks upon his face and limbs, as the bitter winds have left the history of their passing upon the Mill.

The history of the Wellington Mill is a history of hard, and, as the world views the necessities of existence, unromantic, toil. It has stood there since 1815, the year of Waterloo, and its name, cut in grey stone and set in the brick base, is a humble compliment to the Duke. Humble enough at the time, doubtless; one among a mort of such compliments; yet lasting longer, and certainly more worthy, than the glitter and display, the grand dinners that were ate, the rare wines that were drunk when all England celebrated the final defeat of Napoleon. It was a compliment after the true manner of labouring men: at once a thanksgiving that the great ones of the earth had returned to sanity and a brave endeavour to grind wheat for a hungry world.

At that time the Wellington Mill was a conspicuous landmark on the road between East Ham and Barking. Its owner prospered in spite of competition with the Abbey watermill, at Barking Creek, which had been worked since the time of Henry VIII, ostensibly for the benefit of the poor. The Mill was a sound, workmanlike affair in those days. Its pitch pine body stood upon a red brick base, and its cap was of Spanish chestnut. It measured seventy feet from base to cap, and the sails had a span of thirteen yards.

It worked with four pairs of stones: three pairs for wheat, and one pair for cattle corn. The flour was sent all over London, to bakers in the city and to fashionable pastrycooks in the West End.

The Mill has remained in the same family since 1815, and the old man regards its skeleton with much the same feeling as he might regard the bones of an ancestor. And indeed it is an ancestor. For the fortunes of his family have depended upon its cunning sails to defeat the caprices of the wind. He told me of anxious days of longing for so much as a "breath," and of other days when some mad god juggled with the earth on the end of his blowpipe. But, wind or no wind, there was always work to be done: stones to be dressed, sails to be repaired, corn to be drawn up and stored in the bins.

His earliest recollections are of the Mill and work in the

Mill, of aching limbs and gashed fingers and, in particular, of the only tragedy connected with the Mill.

It happened one gusty Whit Tuesday, seventy-four years ago. The wind yelped about the countryside like a pack of angry dogs; it snarled about the Mill, rattling the pitch pine scantlings and snagging at the door-latches; it snapped and retreated, leaving a damaged sail. A labourer was sent out to repair the breakage and, as he was firmly astride the vane, the wind returned with such ferocity that the propeller broke, man and sails falling sixty-six feet to the earth. The man was dead, the sails were shattered to matchwood.

New sails were fitted and the Mill ground out the grain. It worked until its timbers groaned against the domination of the winds. But, like men themselves, it was in the service of man, and it was patched and strengthened, and it ground on. It worked until the sails fell to decay and the moonlight streamed between the gaping and rotting timbers. The air raids almost completed the work that Time had begun. Bombs fell frequently in the district, shivering the old timbers and causing fresh gaps in its side. On these dreadful nights the old couple sought refuge in the Mill and stayed there until the danger was past. He told me this with many pauses and whippings of memory; and as I rose to go, he said without bitterness that the days of windmills are over. His Mill is too old to repair; and the corn is ground by large concerns that can maintain a steady output. His old wife, bent and wrinkled, touched my arm and said: "Yes; but we cut a little chaff."

To cut a little chaff.

That, at least, is one way of cultivating one's garden.

THE MILL, RAMSGATE

LAST OF RAMSGATE MILL

1

WHOEVER goes tilting windmills must be prepared for surprises. One expects to find many beautiful windmills fallen in decay; for, except in rare instances, there is no niche for them in this world of steam and oil, neither is there, so far as I know, one of those amazing societies with a suite of plush offices whose business it is to restore and protect them. And perhaps the reason for this neglect may be found in the fact that windmills are in nature essentially democratic. The old castles have their seneschals, the old forests their keepers, supported by moneys that are always forthcoming when any aristocratic hulk, left stranded by the sea of Time, is in danger of falling to pieces. But windmills are the old labouring men; and mankind in general is not very much interested in labourers who are too old to work, or whose labour is now a homely antiquity. So these tired and broken grinders of grain are left to rot in ignoble fashion about the countryside, assailed by winds that were their souls aforetime, and mocked by gorping stars that once danced high, inscrutable dances between their whirling sails. But, in one or two instances, humiliations more cruel than thoughtless neglect are practised upon their bones.

At Ramsgate, in Kent, the people are kindly after the true manner of Kentish folk. In spite of the fact that they tolerate the tawdry attractions cultivated for the delectation of trippers, they are unspoilt, and, standing as they do with one foot in the sea and one foot in the most beautiful county in England, essentially simple and wise.

They had a windmill—once; a lovely affair whose frail lattice sails were visible to homeward bound sailormen approaching the harbour. For nearly three hundred years it was a joyful landmark for sea-weary eyes, that windmill above the sea with sails circling against the clouds and the blue. Now only the topmost yard of a headless body can be seen, and perhaps, if they are close in and use a good glass, patches of divers-coloured posters advertising matches, and dog-cakes, and God knows what! They are, mercifully, spared the rest.

The kindly people of Ramsgate, however, see the shameful rest when they pass along Grange Road; they see it with half-averted eyes, I imagine. For that shameful rest is a motor garage!

2

As I walked into the garage I went hot and cold by turns. A nice boy with all the grease in the world upon his hands and arms and face pleasantly asked my business. I hardly knew my business. I just felt as if I wanted to punch the garage to pulp. . .

I had arrived in Ramsgate weary and adust, but with hopes running high. I had come to see the Mill. A hot, wet sun spattered everything with gold; flakes of fire danced upon the sea; the brown sails of the fishing boats were rubied by the fierce sun. On the stalls along the "front" little platefuls of yellow cockles seemed to lick up the light, and the pink prawns kicked innumerable legs towards the sky.

As I was toiling up the steep towards the Mill, a party of trippers passed by. One was singing.

"*Do I want to see my mother any more?*
Do I?"

And a mighty chorus answered.

"*Yes, I do!*"

I wanted to see the Mill . . .

And now I stood in a motor garage, and a nice boy was pleasantly asking my business!

I answered in a shamefaced way that I was interested in windmills. His gaze wandered from the neatly-arranged row of motor cars to the ceiling.

"Oh, yes, the Mill."

We walked outside.

"It was a mill—once . . . but now you see." He pointed to a sign which read: Scott's Motor Garage.

"Quite," I said affably. "It's very quaint.'

I was really too amazed to enter into rational conversation, and I stared stupidly at the single floor garage above which the black, decapitated body of the Mill merged skyward.

"And the miller; where is the miller?" I asked after a pause.

The nice boy leaned against a scarlet hydraulic petrol pump on the pavement and said:

"Miller? . . . Oh, perhaps you would like to see the guvnor?"

I nodded; and as I was crossing to the other side of the road, in order to get a full view of the windmotorgarmill, an electric tram missed me by an inch or so.

The "guvnor" arrived in his car.

"Interested in windmills?" he began. "So am I. Fine old place, this! I am having a wireless set installed next week."

A wireless set!

"Yes, a fine old place," he repeated, "gone to rack and ruin. You see I've used part of the sails, there—that post—they were lying up above, rotting, doing nothing."

He was referring to a post to which a motor-cycle was chained.

He told me that in wartime he used the Mill as a "dug-out" during the frequent air raids on Ramsgate, and that one "got a fine view through the open top."

I regarded him with admiration, constrasting his hard, classical view with my own pulpy, romantic vision.

He is the new "miller."

I shook his hand heartily.

3

The Ramsgate Mill is three hundred years old, and it stood originally were the S. E. & C. railway station now is. About one hundred years ago it was moved and re-erected towards the Pegwell Bay end of the town, and there it has remained. In those days the surrounding land showed a sea of billows as green and as restless as the Channel billows below; and the Mill ground the wheat and corn for the town and country-side. It worked, with many restorations, until twelve years ago, but the big business had gone long since, and during its last years the Mill trade was almost entirely confined to supplying small quantities of flour to housewives.

Dry-rot set in; the sails were removed; and then the cap and a fair part of the body were torn off one stormy night; the stones were sold and sent to India for rice-grinding.

Originally the Mill had seven floors, with a pair of stones on each floor. Three floors remain, so rotten and damaged that ascent is fraught with broken bones. I took the risk and escaped with nothing worse than a sad heart. All is chaos within: splintered floors, pieces of sails, smashed bins and fragments of stones.

I climbed to the top of the latter and looked out to sea. Far on the skyline I saw a large vessel followed by a dense cloud of smoke. Below, in the garage, the roar of an internal combustion engine cut my brain as a sharp reed cuts the finger.

I looked at the wreckage about me and——

Well, it is an old story.

THE MILL, WINCHELSEA

WINCHELSEA MILL

N the road to Winchelsea I met an old man, bent with toil, who moved as if heavy weights were attached to his feet, but his heart was light as a bird's and his eyes clear as summer. I told him how many miles I had come to see the windmill at Winchelsea, and I asked him what he knew about the windmills in those parts. He eyed me suspiciously, and, after a full minute's silence, said: "You're a rum 'un." Even he, old man that he was, belonged to a newer age.

The town of Rye, where Henry James wrought his black magic upon the language, lay behind us, and we were walking by a stream along whose banks willows drooped like sad sentinels. Straight ahead, high up, was Winchelsea; and, as we turned with the stream, I saw my Mill on the crest of a fine windy height.

"That's it," I said, pointing.

"That's a dead 'n," he replied. "Blown to bits."

I rated him for an unimaginative churl. He answered that mills were nothing to him: *he* remembered the first steam plough in the district. That was a tidy few years ago. He told me that its owner was a tall, thin man who drank quarts of raspberry vinegar, carried on with four women at the same time, and "studied himself into the grave." He was, I gathered, habitually clean-mouthed, but he was always accompanied by a man who did the swearing—a man who wore a leather boot-lace as a watch chain. I parted with the old fellow at the foot of the steep. His business—he was on a visit to a niece suffering from mumps—took him to the left, while I turned into the fields on the right.

After a stiff climb to the crest, I called upon a farmer on whose ground the Mill stood. He was a new-comer, and, judging by his accent, a man from the shires. The Mill meant nothing to him—it was just lumber: he was interested in sheep. He asked me what I could see in Mills, *nowadays.* The last word was stressed pityingly.

"I find them charming," I replied, "as some people find armour, and postage stamps, and birds' eggs charming... I may even write a book on windmills some day..."

"Some folks read queer stuff," he dryly remarked.

As we walked over to the Mill, he told me an interminable story of a dead body being found near the place, and how a furious dispute arose between the coroners of Winchelsea and St. Leonards as to which of them should conduct the inquest; for half of the Mill is in one parish and half in the other.

The Mill is built on frail, feminine lines. It reminded me of a respectable old woman whose dress has been patched and mended until very little of the original material is left. It was built two hundred and fifty years ago by the Grey Friars for the purpose of supplying the monastries with flour.

Until twenty years ago the Mill ground corn and wheat for the village corn-chandler's shop; but its structure fell into decrepitude and, although it was strengthened many times, it could not be worked profitably.

In spite of the great strength of the propeller, which, I am told, weighed more than a ton, the sails were blown off. Since that time the Mill has been left to rot; indeed, I could not find out who owned it; nobody knew or cared. And so it stands there on the hill, beautiful in decay, at the mercy of the rats and the weather.

The land falls away on all sides. Northward the country is divided into coloured, cultivated squares, with here and there farmhouses and groups of cottages, and in the far south a thin blue line of sea melts into the sky. It was so quiet that I sat there until the earth was all a darkness, and the merest shaving of a moon was drifting into the night. Impulsive winds played about the Mill and passed on to the sea, hissing like sudden surges through the trees.

O Mill, I thought, your fate is very like that of man. You worked blindly and towards an unknown end; but *your* end was certain. We also work on blindly, knowing next to nothing, guessing much, hoping that we may be respited even as the darkness closes round us . . .

My pipe was out and, rising a little regretfully, I walked towards the winking lights of the village.

THE MILL, LITTLEHAMPTON

ITTLEHAMPTON by the sea is one of those forced growths which are so assiduously cultivated by enterprising railway companies. There is a lot of sand there, and a lot of sea; there is a sky of changing colours, in which our benevolent Mother Nature occasionally hangs a sun. All these things are good for commercial enterprise—they are fine material out of which are evolved bright posters for the walls of our cities.

But posters do not make a town; and Littlehampton is not yet a town. It is just a collection of boarding-houses and, a little farther inland, of a few decrepit agricultural dwellings. And it is still primarily an agricultural hamlet, whatever the posters tell us, and in spite of the fashionable dresses that during the summer season flaunt their gay colours to the day.

Let us forget the posters and the boarding-houses and the bright dresses, and let us look at the old Littlehampton, a dark village set in grey sand-hills against which a grey sea like shrunken flannel beats unceasingly.

There is an old wooden pier, one of the oldest in the country, a noble structure which the white teeth of the foam have bitten in vain; for it still stands four-square and triumphant.

And at the end of this pier, looming monstrously black against the low horizon, is the windmill for which I had relinquished the modern Littlehampton with its pleasant frivolities and its happy summer children.

As is my custom, I sought out the nearest son of the soil and began to talk about the Mill.

"It's a gonner," said he, "but its been a good 'un in its time."
I looked at it and nodded assent.

"I remember," he continued, "when this old Mill ground nearly all the stuff we wanted; but those days are gone, and now it comes from other parts."

"How old is the Mill?" I asked.

"Two hundred years, or thereabouts; but I'm not sure."

Windmills are curiously like human beings. They have each a character of their own—or a lack of character: they are sad, gay, robust, frail, masculine and feminine. This mill at Littlehampton is a strong forceful personality, tragic in defeat. There is about it a warrior air, a subtle something that speaks of battles and forced marches, of melancholy bivouacs under stormy moons; it is marked by unceasing travail and by fierce joys that leave scars like sword-cuts. The Wellington Mill at Barking, of which I have written in another part of this book, has a similar character, but it is a heavy dragoon, while this Mill is a marine. It has fought battles both on sea and on land.

"Is it not strange," I said, for I suspected that the man had a soft spot in his heart for the battered hulk before us, "that nobody seems to care what becomes of these old mills?"

"Well, sir, it is and it isn't," he replied. "You see it is like this: windmills don't pay; there's nothing in them; and they are a rare lot of trouble too. Hard work and not much at the end of it. But they are fine and grand, all the same. Now, if this Mill belonged to me, I'd have her painted up and put a few flowers in boxes round her; it would do her looks a world of good... But it isn't mine... It will fall to bits one of these days."

"You've a few flower beds of your own, I expect?"

"Yes, I rear a few flowers," he replied. "It's my hobby. And the missis like 'em."

He was a fine upstanding fellow with a face the colour of mahogany, and huge hands, giant-veined, flattened and broken nails—superb natural vases for his flowers.

"Children?" I asked.

"Two; both boys . . . rips." He smiled with pride.

"Do you remember the miller," I asked, when the softness had gone from his eyes.

"Yes; dead and gone."

The mill, too, is dying and will soon be gone. It is a pity: but then, as my friend said, there's nothing in them. He loves windmills almost as much as I, but he has other loves, stronger, more vital loves. So have I. So have we all.

THE WINDMILLS: CAMPO DE CRIJITANO

THE MILLS OF CRIJITANO

AT Campo de Crijitano, a Manchegan village in Don Quixote's own country, there is a collection of windmills such as will quicken the pulses of all lovers of the worthy knight. They are etched on my brain in clear, fantastic lines, and yet the picture is so unreal as to seem a crazy fragment of an opium dream.

Campo de Crijitano, being outside railway communications, has preserved all the ancient and picturesque qualities of La Mancha. The old houses, the old customs and dresses have remained unaltered by the march of time and civilization, and as one walks into the village one leaves the centuries behind. It has all the old discomforts and old pleasures. The single dangerously steep street is as it was ages ago; one strolls into the same posada, one drinks the same wine out of the same skin, and the people talk of the same matters in precisely the same crystallized sentences as they did when the knight passed through the village to see Dulcinea the beautiful transformed by black magic into a country trollop.

The land around is rich and fertile, and life flows between prosperous banks; but prosperity has in nowise altered the standard of living or caused a deep division between the classes such as we see in England. The good God has ordained that the soil shall be fertile, therefore thanks are due to the good God. But, for the rest . . . well, life moves as it always has moved, happily, on the whole, certainly smoothly. Why should the poor grumble or the rich grind? They are one family. . . They are wise, these Manchegans, and they have compressed their wisdom into proverbs which will stand as long as Spanish is spoken—proverbs which point continually to lines of least resistance and the need for work and a little tolerance. And the Manchegan works twenty hours out of twenty-four . . . The lazy Spaniard!

The windmills stand on a gentle height just outside the village. There are a dozen or more of them, plain, plaster-thatched structures, supported from behind by saplings; the sails are enormous and beautifully curved, and, seen from a distance, they cut the quivering blue sky into a fantastic mosaic.

A man and a donkey were treading their ways up the slope.

"It is hot," I said.

"Yea, brother, it is the sun."

I craved a drink of water and observed him closely. He was a magnificent fellow, six feet high with limbs and a body that instantly lounged into a voluptuous repose. A coloured kerchief covered his head, the face below was like a copper shield, and the black eyes were full of sunlight.

I offered him a cigarette.

"*Bueno,* but it is good to smoke."

The countryside drowsed, drenched in sun and fanned by warm winds. Near the horizon a solitary white plume of cloud drifted across the blue towards the thin, faint line of sierras.

"It is better to be rich," said I.

"A good friend is better than either," came the easy answer.

"These windmills are very friendly to man," I said. "They are also very old, and what is old is good."

"Yea, but one cannot call them brothers. They are like this mule," he replied, betraying at once the typical Spanish indifference to animals. "Old? How old I cannot say, and *quien mas miente, medra mas*. . ."

"But, they work for us; therefore they are our friends."

He stared at me with a frank insolence which contrasted strangely with his puzzled features.

"God has so willed it." And he crossed himself.

The fellow was extraordinarily pious; his mind was a perfect museum of theological calashes and strait jackets. I probed him gently and discovered beneath his courteous manner an adamantine intolerance. He really cared for nothing but the most threadbare symbols of his creed: its underlying kindliness and glorification of the soul of man left him cold. Sensuous worship—just that; a habit of mind (I can hardly call it anything else) which harmonizes admirably with the landscape. For there are no half-tones in Spain: everything stands out clear and stark. And the hard physical outlines kill whatever mysticism there may be in the Spanish character. Blood and flesh are life—and death?—flesh and blood. A London fog might easily transform these people into something approximating to Unitarianism. . . Hardly that, perhaps, but something on those lines. But, God forbid! The world has need of these callous Spanish physical values.

This man was quite exceptional, for very few Manchegans —the men, I mean—take the worship of "Our Lady" seriously. They seldom attend the churches; that is left for the womenfolk; they prefer to slack from their arduous labours, uttering savage blasphemies the while. I have been told that "the priests make a good thing out of it, while the people work to keep them." It may or may not be so, but that is the general feeling.

We parted with a "Go your way with God," and I was left alone with the Mills. Their numbers overpowered me. There is something not quite decent in putting so many windmills together; and it is difficult to focus more than one at a time. A single windmill intrigues one by its solitary beauty, but when one sees a dozen or more slouching together like drunken giants—well, who sees the charm of a single leaf in the forest? Small wonder that the noble knight mistook these windmills for a brood of giants!

I see them now, savage and sombre as the people for whom they grind. And whenever I think of these windmills at Campo de Crijitano there is a deep purple sky over the earth and the air is full of soft hummings like the plash of distant waters; mule-bells shower intermittent cascades of cold silver music; a dog barks; a man's rich voice bursts impulsively into song and ceases, and his footfalls grow fainter.

I stare into the night and my thoughts go along the road to Andalusia, for there—but that is a personal matter, and I will spare a long-suffering public. In any case, I should probably be accused of sentimentality or bad form if I mentioned it. So let it pass.

THE MILL, KESTON

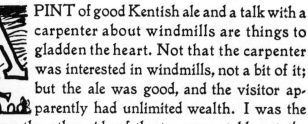A PINT of good Kentish ale and a talk with a carpenter about windmills are things to gladden the heart. Not that the carpenter was interested in windmills, not a bit of it; but the ale was good, and the visitor apparently had unlimited wealth. I was the visitor; and on the other side of the tap-room table sat the insatiable carpenter.

He had a roving eye and a knack of imparting the dullest commonplaces with an air of sacerdotal authority. Perhaps he was a good carpenter. I don't know... But if drinking is the sign-manual—well, this fellow had a throat like Nebuchadnezzar's fiery furnace.

It is all very well for moralists to talk about the evils of drink. It is easy to talk about the evils of anything; but, as far as I can see, morals have nothing to do with windmills and carpenters. Certainly nothing to do with the windmill at Keston in Kent and the carpenter who sat opposite to me. He was as unmoral as the Mill was featureless. He was just a liar with a thirst; but, whether his lies were the outcome of his thirst or vice versa, I am unable to determine. It would be worth a moralist's while to call upon the carpenter, for I am sure that many learned societies would welcome a monograph on The Carpenter of Keston... But I advise him to keep a tight hold on his money...

After his second pint the carpenter jerked his thumb windmillward and, after expectorating, said:

"There's a good bit of timber in that Mill."

The truth of his observation being manifest, I knitted my brow and acquiesced solemnly, and waited. I felt sure that there was more to come. He braced his shoulders and literally bit at the ale, and lo! I saw the shining end of his nose through the clear bottom of an empty glass. He looked at me triumphantly... And then at the bar...

"Yes, a goodish bit of timber..." his voice failed. His eyes probed my soul. I suggested another drink.

"It does me good," he said simply.

He produced a large empty pipe and, after blowing through it violently, made a movement as if to return it to his pocket. It stopped midway on its journey; he knew me for a man of understanding. I made no attempt to resist his ruffianly purpose.

"I do enjoy a bit of real good bacca," he said confidentially, transferring a handful to his inner pocket; "but trade's bad and it don't run to it." He sighed and returned my pouch—empty.

"The Mill?.." I began.

"Oh, yes; well, you can see for yourself that there's a goodish bit of timber in her... a goodish bit."

"That's all right—the timber," I said somewhat tartly, "but do you know—"

"Ah, it's like this. That Mill's been there a goodish few years—why, when I was a boy that height," he indicated about two feet from the ground, "about that height, I know Mr.——"

"How old is the Mill?" I demanded. "How many years ago was it built?"

"Why, now I come to think of it . . . well, more than a little."

He emptied his glass. Obviously, there was nothing to be learned by using shock tactics. I must wait patiently and let his shy soul find its wings.

"Once when I was on the oil——" he began.

"The *oil?*" I queried.

"Yes, oil—this stuff," he lifted up his empty glass.

"Oh!"

Another pint.

He was beginning to like me. His roving, lambent eyes were full of approval, and, bending over the table, he "tipped me the winner of the big 'un." I regarded him stonily; and mistaking my coldness for scepticism, he whispered behind his hand:

"It's the goods, old chap. Put your shirt on!"

Old chap! My shirt!

My God!

Rural England!

I thanked him with hate in my heart. He must have glimpsed irony in my eyes, for he began:

"The Mill. Some say it's over a thousand years old and that it was built by Queen Elizabeth after the Great Fire of London. Mind, I'm not saying that that's Gospel. But some say so. . . Of course, I'm a carpenter, and I know a good bit of timber when I see it. The Mill's a good bit of pitch-pine. . . A brother of mine could have told you all about it—he worked thereabouts when he was a boy; but he's in Canada now—Moose Jaw—and his wife's just had a nipper. I got a photo last week... You see after the old man died I went on the oil for a bit."

Another pint.

He could see that I was not impressed, and he made a desperate effort to retrieve his position as an oracle.

"Of course, the Mill has done a fair bit of grinding in its time, a fair bit. You can take that from me. Sacks and sacks. I've seen carts loaded up day after day—thousands of 'em. It's done a fair bit of work, I can tell you. But I'm a carpenter. That's my job. But I know a good bit of bacca when I smoke it."

His pipe was out.

My pouch was empty, and I made no sign of having heard.

"But, as I was saying, I'm a carpenter. Perhaps if you stepped across to the Mill you might hear something else."

I didn't step across.

I had heard enough.

THE MILLS, BRUGES

FURNES, BELGIUM

THERE are so many windmills around Bruges that it would need a volume to give an account of them, and as they are so much alike, both in spirit and construction, it would be tedious to attempt anything approaching a detailed study of them. Bruges and its environs might indeed be called Windmill Land, for there windmills are as common as pawnshops in Pimlico. But they are none the less beautiful (the windmills, I mean), and I have no doubt that each windmill deserves to be treated separately, but I would as lief confine my attention to a single blade of grass in a meadow as make a preference of a windmill in Bruges. Besides, I don't feel in the humour... which is perhaps the real reason.

No; I like the massed effect, the high blue sky and the deep level waters, the fantastic willows and the hooded women washing, the dark reflections and the slow-circling sails. There is nowhere in the world a cleaner picture than this Windmill Land which looks as if it has been soaped and scrubbed and dried and polished by a million industrious housewives. I feel as much at home there as in my own kitchen; I can lounge and stretch and smoke my pipe, and I am just as careful with my tobacco ash. It is a place of decent, homely peace, such as all restless men desire, here on the earth or in a distant corner of the sky.

I know that I shall never visit Windmill Land again. I do not know what changes may have taken place there during and after the Great War, and I have no desire to know; for me it is just a quiet bourn where I can retire after the heat of the day, where I can walk along the river edge in the cool of the evening and see the long line of windmills wind into night with the waters... The Great War? Historians have bottled the smoke which floated high above Windmill Land —bottled and labelled it for the dark museums of man's madness. Doubtless the last breaths of the finest are mingled with that smoke, and it may be that the historian's bottles are full of human virtue. Perhaps Windmill Land bears outward and visible scars in common with many other lands. I don't know. I shall never know. For I shall never go there again, neither shall I visit the museums of the historians. Windmill Land shall remain for me a rare memory—a quiet place where I can walk undisturbed in the twilight by the windmills and the willows and the waters...

Somewhere in Windmill Land there was a café where I was in the habit of breakfasting and reading the newspapers. The morning sun shone through the window and played among the many-coloured bottles on the shelves, etherealizing the quiet, pale-faced girl who waited on me and flecking the large, polished face of Madame with high lights. I would wish both "Bon jour"; the girl would return the wish in a shy, civil manner and Madame would thunder an embellished compliment in a pulpy contralto. She had had three husbands and eleven children, and her heart was still full of love; she would tell the scandal of the previous day, in a manner now kindly, now indignant, but never bitter. She knew nothing of the world outside her café, and, like all simple folk, she had a keen nose for a tale of passion. I remember how she listened with moist eyes to the story of *Aucassin et Nicolette*, which I related to her while she was washing the glasses. But, when I talked of the romance of windmills, she laughed and called me a *farceur:* they were a part of her life and she had hardly noticed them.

After breakfast I would walk along the river by the windmills and talk to the labouring men and women. They were all like the good Madame, practical and simple, hardworking and romantic—human editions of the windmills. One evening I met and talked with a priest, fat and kindly after the fashion of his race, and, I suspect, much like his God, except, perhaps, that He is bigger and laughs louder. We talked of the windmills.

"Some are old and might tell much, could they but talk," he said.

"Your penitents are just as old, and, happily, they can talk," I replied.

"That is metaphysical. Many of them drink too much, and at such times the power of speech is to be regretted."

"It is a happy place, this Windmill Land, is it not? The people are poor, but they have nests in this quiet place and are content."

"It is as they make it. Their hearts are sound and they work hard."

We walked along the riverside into the luminous dusk, he talking of the superiority of English cigarettes, while I extolled the virtues of his people. He had a cousin in London—at

Stamford Hill, I think—and he asked me to tell him all about the place. I did so, with my eyes on the dark waters and the windmills, until the land was all a purple peace.

Mill of Lauragais

A PUMPING MILL, HOLLAND

WEERDT, HOLLAND

OUT of the numberless windmills which are scattered over the Dutch countryside, I shall choose two at Weerdt, in the province of Limburg, for the subject of a few general remarks. They are not the most beautiful windmills in Holland, neither are they the most romantic; there is nothing to distinguish them from the thousand and one other windmills which lend the Dutch landscape such a homely charm; but for me, at least, they are at once typical and symbolic. For all I know they may have a pretty history, and technically they may be worthy of special remark: but I only think of them as I think of ordinary cows in undistinguished meadows, beautiful and wonderful as all common things are.

Have you ever stopped short before a perfectly trite piece of landscape and exclaimed, "That's fine," and stared long and long, without being able to discover what is fine about it? Perhaps you haven't; but that is precisely what I did when I first saw the Mill outside Weerdt; and I cannot for the life of me understand why I should have done so.

It was an awful day. A penetrating wet mist hung about the countryside, and the trees and the houses looked like dirty bubbles floating in the murk. I was wet and cold and hungry, and I had sought the famous Dutch hospitality in vain. Such clods! Without imagination to understand the plight of a traveller, weary and footsore, and speaking less than a dozen words of Dutch. As I stared at the flat, pitiless grey land I could have wept with rage and disgust, and I sat on the soaked bole of a tree to exploit new combinations of swear words, and to spit at the memory of lying calendars and posters which had played no small part in inducing me to come to Holland.

"Damn it," I muttered, "if I could only see a pair of baggy trousers!"

And suddenly the dark form of the Mill, a mere stain in the grey, crept into my vision.

"That's fine!" I said.

I forgot all about the two miles walk that lay between me and the town; I forgot that I was wet and cold and hungry: this shadow seemed to me more vital than any of the discomforts which assailed me; this mere ghost... Goodness knows, it looked mysterious enough and intangible enough to fade away at any moment; but it held me as strongly as if it were a newly descried castle in a grey forest, and I sat and speculated upon the wonders it contained... And then the calendars and the posters came into their own.

I saw in the mist a circle of fat Dutch housewives treading an elephantine dance to the music of a tin-whistle played by a boy who sat on the cap of the windmill. There was no joy in their movements; they were intent, serious, stupid; and as they passed me one by one, I saw that they bore printed cards on their backs, advertising somebody's mustard, beer, pickles, cheese, vinegar, sardines, soap, cigars, gin, milk, fuller's earth, matches, safety-pins, mouth-organs and countless other commodities with which the wily merchant kills the desire for suicide in the human species. When the last fat woman had passed, the circle was again formed, this time by men. They all wore floppy caps with shiny peaks, baggy trousers and sabots, and all smoked long pipes. They performed precisely the same movements as the women, and each man carried an advertisement on his back. I jumped forward and tried to break the circle in order to enter the windmill, but they seized me and stuffed my mouth with their vile wares, pricked me with pins, tickled me with lighted matches and aimed huge red cheeses at my head. The mist engulfed them, and I felt myself falling rapidly to earth...

I awoke with a start, shivering violently. The mist had thinned, and a weak sun dribbled anæmic beams out of the wrack of grey sky. The Mill shone darkly before me; the sails fluttered like the wings of a dying bird; and a perfectly ordinary man was backing a cart towards it.

I stamped the ground under my feet and walked towards him. I won't attempt to describe our efforts at conversation, but after a time the fellow grew tired of me, squatted on the wheel of his cart and opened his lunch packet.

It contained cheese.

I thought it was about time to move, and I began my journey to the town.

      *   *   *

On arriving in Weerdt I learned that the place was *en fête* for the day; on what pretext I do not remember. After a meal and a change of clothes, I made my way through the bright holiday crowd towards the centre of the town. I couldn't enter into the fun of the people; the gay flirts of gesture and the shouted banter were lost upon me: I sauntered leisurely on, feeling very English and a trifle Olympian, and eyeing the folk with priestly mansuetude.

My attention was drawn to two old men who sat in the darkness of a doorway. They were very, very ancient, nearing a century, I should say; their wrinkled faces had the texture of rotting cherries; their limbs were crossed in crude angles, for the years had filched the rhythm of their forms. They took no interest in the moving crowd. Their eyes were fixed on I know not what eternal things; and I watched them intently for some sign of animation, of human feeling, of sympathy with each other, even. A young girl waved a light hand before them, and dropped a gay cascade of laughter: they did not even blink. And I passed on, unable to say definitely whether they were alive or dead.

I eventually came to a square in the middle of the town where a large crowd was collected round the base of one of the most beautiful windmills I have ever seen. Its sails seemed to fill the sky like the wings of an enormous bird, dwarfing the church and the houses, and holding, as it were, the noisy multitude in a maternal embrace. It was the great reality, an eternal call to labour, a rock against which the swift joy of the people broke in a fuming spray. So frail, the delicate, scaled body poised perilously on the roof of a barn-like structure, and yet so subtly strong!

Above the shuffle of passing feet I could hear the mighty groaning and straining of the still sails under the assaults of a persistent wind, and the dampness, on which the sun struck bobbing lights, looked like the sweat of a giant nearing exhaustion.

The town-folk kept up the merry dances, the joyful cries. The Mill alone still laboured, speaking to me with a voice more powerful than any, calling my mind to the necessities of existence above the faint, far voices of human mirth.

THE MILL, HERNE

A KENTISH MILL

WHEN a windmill falls into commercial hands there is very little likelihood of its retaining anything of its pristine charm. It becomes a factory; it has fixed hours of work; and when the wind has a lazy fit and hides from the traps laid by man an electric dynamo is brought into operation, and the work goes on. It is a great pity that any windmill should be brought to such a pass, but the economic pressure of modern life respects nothing out of which profit can be drawn. And so, here and there, one finds a windmill that is being run as a serious commercial proposition.

Such a mill is that at the village of Herne, near Herne Bay, in Kent. I descended upon the place eager to see a mill actually in working order, to watch it in operation and to have a chat with a real live miller. The Mill is certainly in working order, and the miller is real enough and live enough. But as for the chat I had so eagerly anticipated—well, the miller treated me as if I were the representative of a firm that sold poisonous oil-cake or an income-tax collector. The atmosphere of the Mill was so thoroughly commercial that instinctively I fumbled for a card. The miller did not wait for it. He had no time: he told me that he was very busy, but if I cared to wait he would be pleased to show me over the place. So I waited. . .

The Mill stands in a fair stretch of ground, and scattered about here and there are one or two cottages and a farmhouse. Even the farmhouse smelt of commerce; it was the sort of place a company promotor would run as a curative exercise to reduce his waist-line. But the land around is beautiful and typically Kentish. The sky was quite uncommercial, and the green fields were as removed from the spirit of Throgmorton Street, E.C., as is heaven.

I waited and waited for the miller to return. I walked round and round the Mill, until my mind became a blank. . . There was a lot of machinery—thoroughly up-to-date stuff, on which I was able to distinguish the patent marks; there was a Ford car which was being loaded up with sacks of flour; a gasoline engine which supplied power for the machinery; the debris of a mechanic's shop—spanners, nuts, bolts, and empty petrol tins. . .

I waited so long that I was able to take a complete inventory of the place; and still no miller returned. To calm my impatience I began to count the number of paces it took to make a complete journey round the Mill, and, finally, the number of journeys I made. It is, I suppose, a form of madness, but there was nothing else to do, and the Mill bored me to extinction. I sat down and lit my pipe, and worked out a simple arithmetical problem. After a terrible mental crisis I discovered that I had made eight thousand, six hundred and twenty three paces in my journeys round the Mill; but I am a precise man and like round figures, so I arose and again began to walk round the Mill until I had completed the remaining three hundred and seventy seven paces, making a total of nine thousand.

Still no miller returned.

So I walked off fully determined to complete ten thousand paces on my way to Herne Bay.

THE MILL, WHITSTABLE

ONE sees the Mill at Whitstable on a height above the town, a graceful old lady, sheltered from the cares of the world, quietly knitting stockings with four giant pins. She is all that honourable peace is, and all that rest after heavy labour is; for after one hundred years of toil she is enjoying a serene old age under the kindly care of an artist.

I wanted to chat with her and to put a nosegay in her lap, so I climbed up the winding road towards the hill, and, after gazing at her for a while, sat down beside her chair.

"I hope you are well," I began.

"Yes, I am getting better every day," Mrs. Mill replied. "But perhaps I ought not to say that. One of my children has been reading the newspapers to me, and I learn that a dreadful person called Coué is telling the world that if folks would say what I have just said they would live for ever. And I don't want to live for ever. Just a few years more to repay the kindness I have had shown to me."

"You are very happy now?" I asked. "Happier than you were in the old days?"

"Happy—yes; but not happier than I was a hundred years ago. You see I was too busy to be unhappy then: I worked night and day to fill the mouths that were dependent on my labours. Now I have much leisure and loving care—more time for thought, and at times, thinking over the past, I long for the old days when my arms were arms indeed."

"But," I said, "if you were as strong as you were a hundred

years ago, there would be very little work for you to do—and I am sure that you would not like to be used as Mrs. Mill at Herne is being used."

"Oh, what is the matter with Mrs. Mill at Herne?" she asked a little stiffly.

"I called on her the other day," I said. "She is in a terrible plight. All her people are dead, and a stranger owns her and is driving her by an engine. She was too unhappy to talk to me."

"Yes, it is very sad; but I should never be treated like that. A woman of her sort can hardly expect anything else. She was always lazy and unreliable, and she dressed shockingly. You saw her hat! Such a slattern... My men have always been proud of me..."

"You certainly look beautiful now," I said. "I see you have a new tar coat, and your flowers are lovely."

"Yes, my old coat let in the rain, so they bought me this," she proudly answered. "And they have built a conservatory in my side... There were few flowers in the old days, I am afraid, two flours only. Oh, how silly of me—I didn't mean to say a stupid thing like that!.. I was very old and near to death when my new master came, but he strengthened my weak parts and cleaned me out. Then he built the rooms and came to live in me. I was so pleased; and I expected that I should be set to work again. But no: he removed my stones and turned my sails into knitting pins, so I just knit a little now and then, for I must do something, you know."

"Well," I said "I am very glad to have seen you again, Mrs.

Mill, and I hope that you will live for a long time yet." And I left her to seek the owner.

He is an artist who bears a name famous in the annals of men. He was unfortunately out, but I was politely told that I might look over the Mill. The four floors have been preserved and converted into living rooms. There is a conservatory with plants and flowers to gladden the eye, and there is everywhere a sense of secluded comfort. There are inviting chairs and bright curtains and a neatness which is admirable. The old grinding stones have gone, but their places have been taken by lithographic stones, and on one floor there is an artist's press.

I thought of many mills I had seen in different parts of the country and I could not recall one which affected me in quite the same way.

Near the Mill are two cottages which seem in some way to be a part of the place. I was told that they had been built within the last few years, but, sitting there, I preferred to think that they were a part of the Mill, or the children of the Mill.

There is absolute peace and solitude on the crest by the Mill, a sense of remoteness from the affairs of men; and I sat down to enjoy the peace and the prospect. The road of approach winds away into a valley, and on the other side the land dips to the sea.

So with heart at ease I watched a high wind tearing a few white fluffy clouds to shreds.

THE MILL, DUNKIRK

THE Mill at Dunkirk is for me fraught with memories of the war days; of long nights of "standing to," the *durump, durump* of German aeroplanes and the sudden shattering explosion of bombs; of shivering hours of dockside activity and of drafts arriving in the rain, the never-ceasing rain; of the unnatural gloom of khaki, and the unnatural outbursts of fierce gaiety amongst the townsfolk. Those unforgettable days! The *ennui*, the horror and, yes, the romance, hidden like a seed-pearl in an ocean of mud, oppress and thrill me with undiminished power whenever I think of the Mill. But it had no particular significance for me in those days—the mere animal effort to "carry on" absorbed one to the exclusion of everything. And so, looking back, the towns one saw, the people one met, the queer twisted debris of men and things appear like fantastic angles of a jigsaw puzzle against the dark background of that intense and blind desire to keep going.

The Mill, like the land, was derelict, a dark towering smudge above the sodden desolate miles of ruin. Inactive, it seemed to be the very epitome of the wastage of war; the sails stilled, the timbers rotting, the bins empty and the stones neglected, while the people cried aloud for bread.

I have many times compared the dim memory of the silent Mill at Dunkirk with the many prosperous mills I have seen at happier times in happier lands, and the comparison has always depressed me. It is a sufficiently eloquent reminder of the blind injustice that seems to be bound inextricably with the lives of all earthly things; and, while there is human flesh and bone rotting as the Mill was rotting, it is difficult to drop even a sigh over a memory that, while a fragment of the most dreadful nightmare that has yet terrified the uneasy minds of men, is yet a thing builded by man; and what man has made man can destroy.

But flesh and blood . . .

ST. BENNETT'S ABBEY MILL

THE MILL, ALDEBURGH

IT was while gazing at the Mill at Aldeburgh that I met my Panurge. I say *my* Panurge, for he was a deciduous model of the original, but he served my purpose for the afternoon. And, after all, I am by no means a Pantagruel, rather a Bringuenarilles, the Swallower of Windmills. He was a merry fellow with the most melancholy appearance in the world. His woebegone aspect reminded me of a bedraggled squirrel in the rain: his long, lank figure was bent like a reed and crowned with inglorious dirty-grey wisps of hair and a face mottled with grog-blossoms. But his eyes and lips were full of smiles, and he had a pretty wit.

Now the Mill at Aldeburgh is just the sort of thing to set the imagination of a Panurge on edge; it is a thoroughly unctuous abortion, a neat bit of hypocritical castration. It was originally a fine masculine affair, strong, austere, frowning—a very buccaneer of a windmill, divorced from human sympathies and fighting a solitary battle against the implacable forces of nature. It gave up the ghost years ago when, worn out by ceaseless toil and struggle, it stood there in black armless majesty, battered into defeat, its sinister face spattered nightly with the blood of dying suns. And if people were possessed of only a general sense of fitness it would have been left in that condition until the final dissolution, for such a condition would have been right, and proper, and in keeping with the spirit of the Mill—that spirit of brooding, revengeful hate against all earthly forces and beings. But, unfortunately, many people haven't that sense; I don't know why; perhaps they are too moral. It is certain that they are too something or other . . . but when it comes to putting a girl's chemise on the body of a helpless pirate—well, it is too damnably, tragically comic. I don't mean to say that somebody has put a chemise on the Mill at Aldeburgh; but that is precisely the effect the "restored" Mill has upon me, and upon Panurge, too, if his remarks meant anything at all. What has actually happened is this: the Mill has been converted into a dwelling and adorned with plaster casts of saints in postures of supplication and sanctitude. The effect is shocking and ludicrous: one reels under the crushing inanity of the thing. I suppose it is the work of pious hands, but——

"O God," I exclaimed, "the *tutti-frutti* buccaneer!"

Panurge heard me, smiled, and roared:

"It's a pæderast!"

"Well," said I, "it is certainly something like that."

"Mind," said Panurge, "I'm not giving that as a strict definition—the saints might object, and I wouldn't offend any man knowingly. I like to be just. '*Tutti-frutti* buccaneer' isn't bad, but it doesn't go deep enough. Visualize it properly, focus it, drink it in, and you will agree, I am sure, that the general effect is that of Priapus after a surgical operation, the nature of which you may be able to guess. My provisional definition was exclamatory and inexact and, I may say, metaphorical, for I hadn't realized its manifest impotence . . . Priapus the eunuch. O the poor wench! . . ."

"Let us hope that the next storm will blow it down. It's disgraceful."

"Rather let us turn disgrace into triumph. Let us make it the next president of the Malthusian League; such a head might inspire men to emulate the example of Attis. By God, it's a great notion!"

I laughed and suggested that he should shew a lead to the world.

"You be damned, sir!" he shouted mirthfully. "I don't count; besides it's too late . . . But I wonder who lives there; probably a seafaring man; some of them 'get religion' very badly. I remember the skipper of a trawler, a hard-swearing, hard-drinking, hard-wenching fellow,'getting religion.' He became amazingly pious and mealy-mouthed, and one really couldn't speak to him. The poor devil got so herring-gutted that we all thought he was starving himself, and, to make matters worse, everything went wrong somehow. First his wife died, then a trawler of which he was part-owner went down, and then his unmarried daughter, a fine wench, had a child by one of the sidesmen or whatever they are called, at the chapel where the old man did his groaning on Sundays. Some of his friends told him gently that it was about time to go out of the god business; but he reminded them of Job and became more pious than ever. When I saw him six months later he was the same old David, reeling about the pavement, blind as a bat. A mutual friend told me the cause of the old man's fall from grace. He went out one night as usual with his trawler, and after cruising about cast his nets, but on drawing them in he discovered that they didn't contain enough fish to feed the crew. It wouldn't have mattered once in a while, but that sort of thing had been going on for weeks; and the old chap was at the end of his tether. Another trawler passed by, laden with fish, and David hailed it and ascertained the latitude of the fruitful waters; and he straight-way steamed to the spot. He then called the crew together and held a prayer-meeting, calling upon the Almighty to fill the nets as He filled those of the Apostles. Then he cast out, and, after the usual wait, drew in and found the nets empty. That was too much, and the old man, seizing an axe, rushed to the bridge and bellowed at the immense waste of grey sky:

'You old ——, if I had you down here I'd break your —— head!' "

"Let us hope," I said, "that this Mill will fall from grace in like fashion."

Panurge nodded and, after genuflecting ironically, accompanied me to the nearest hostelry for refreshment which was, if not Gargantuan in proportion, sufficiently brown to comfort the mind of a Pantagreulist.

THE MILL, SARK

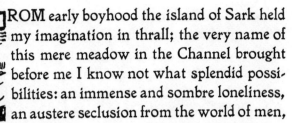FROM early boyhood the island of Sark held my imagination in thrall; the very name of this mere meadow in the Channel brought before me I know not what splendid possibilities: an immense and sombre loneliness, an austere seclusion from the world of men, the magic of wide horizons. I always thought of Sark as a place where a young god might till the earth by day, a peak from which he might touch the stars by night. A silly sort of superstition, this superstition of words, but one which, I suppose, is universal amongst the young. And I am pretty well sure that the fascination that lurks in all sonorous words has driven many young men forth in search of their heart's desire in places like Meleda, Matsu-shima, Timor Laut and Tierra del Fuego. The word "Sark" certainly induced me to go there; and I do not regret my visit.

It has a homely countryside with pleasant roads and byways, and a quiet, plaintive air touches everything with a faint melancholy. But of course the young god of my boyish fancy was not there—luckily, for I suppose he would have been an insufferable bore. Instead, I found the people to be a sturdy agricultural folk, and I spent many happy days during which time I got to know something of their natural kindliness.

There is a beautiful windmill on the island. It stands on the edge of a lonely road, between poplars and hawthorn, and I first saw it darkly bright against the lowering sky. It is neat and workman-like—a plain brick circular body with a tarred cap and most graceful sails and wind-vane.

It has no particular history that I know of, but, as I watched it through the smoke of my pipe, it seemed to me self-sufficient and perfect of its kind: an agricultural poem of consummate fitness, one of those poems, that just *one*, and about which *I* can say but little. And in this respect it is like the island itself and like the people, human and yet cut off from the rest of humanity, living a life in which work and rest after work, a fertile soil and the omnipresent sea form a background against which the destinies of mankind are shown as in microcosm.

THE MILLS, CROWBOROUGH

CROWBOROUGH

TWO hundred years ago some dead and forgotten builder erected a windmill at Crowborough, in Sussex, which still stands foursquare and solid among crumbling cottages of the last decade and villas in the Flemo-Tudo-Georgian style so dear to the heart of the modern architectural stylist. It towers above the pleasant and frivolous modernity of the high street with a cynical grin on its rubicund, weather-beaten face, and one hardly knows whether the black tears of tar which stream from beneath the cap are tears of defiance or despair. There is nothing soft and feminine about this Mill: it is hard and unlovely, and it wins one by its sheer brick force, its barbaric colouring, and its tyrannical domination of the neighbourhood. It stands fully sixty feet high, and thirty feet above the collection of cottages and outhouses grouped around its base. It dims the brightness of an orchard in the rear, and its shadow falls in a duck-pond in the front.

It is, of course, out of action, as nearly all the old windmills are. And I can see no earthly reason why this Mill should not be working: the structure is as firm as the day it was built, and dry-rot, that fell enemy of windmills, has not yet begun its ravages . . . The reason is doubtless an economic one; but that is a hellish reason . . . It ceased work twelve years ago; and many years before that time the sails had been removed and the stones driven by a forty horse-power gas engine, which is still in an outhouse at the base. To facilitate production, I was told; and even that failed to justify the running Mill.

There was a young man mowing a patch of ground on the other side of the duck-pond, and I discovered that he was the owner of the Mill. He was an intelligent fellow who had travelled a good deal, and travelled to some purpose, as I was to learn later. I murmured a few words of admiration of the Mill.

"It's for sale," he replied, "the Mill and the block of cottages; so I suppose it will be pulled down shortly."

"You don't mind?"

"No. Why should I? I've just returned from South Africa—Johannesburg—the climate didn't suit me. I was in the hotel line with my sister, a very good business. Before that I was in Germany on the same game . . ."

His heart was far away from windmills. He didn't even trouble to answer the questions I put to him about the Mill. He was interested in the economic conditions of Spain. Had I been there? I had. How long ago? Three years. What did I think of the chances of an English restaurant in the big towns? Was money plentiful? Was there any feeling of hostility towards Englishmen? And so on . . . Truly, one encounters queer types in rural England. And yet outwardly he was a thorough-going agricultural Sussex man; he had the usual square face with the long, tight upper lip, the same roundness of shoulders and the heavy-footed walk. Impossible to imagine him as *maître d'hôtel*, or even as waiter for the matter of that. He handled the scythe with such a grace as only a true son of agricultural folk can—the slow, deliberate sweep with the body behind it.

"You have no desire to stay here, then?" I queried.

"I should say not! There's nothing here for me. Once a man has seen the places I've seen there's nothing for him in Crowborough . . . Why, there's no *money* here . . . If it weren't for my health ——"

I walked over to look at the Mill from the orchard. Its aspect from behind is even more domineering, and the smashed wind-vane lends it additional force, as a broken nose will add power to the human face. The sunlight glowed fiercely on the red-brick body, lighting the ochre and brown weather-stains and giving the tar a molten appearance, shot with all the colours of the prism, which, from certain angles, looked like congealing blood.

"The giant weeps," I thought.

In one of the outhouses at the back I discovered a bakery, and a couple of bakers were punching the unresisting dough with an earnestness that seemed at once comic and futile.

"Gently, brothers," I thought, "the poor dough can't help the economic conditions."

"I'll bet a pound that your flour wasn't ground in this Mill!" I hailed, hoping to recall them to civilization for a moment.

They didn't even look up; they continued to punch the dough as if consumed by an implacable hatred.

"You win!" one shouted.

"Where do you get it from?"

"Oh, America."

The giant weeps; and no wonder.

※　　　※　　　※

At the end of the town there is another windmill, or, rather, a hideous caricature of one. I don't know how old it is, or whether it was ever worked; nobody knows anything about it. It stands in the grounds of an hotel! It has been stripped and "decorated," and it might be a box that milliners' use for widows' weeds. It is painted rust-black and battlemented; camouflaged windows—black crosses on white paper—are pasted here and there; on each side there is a pair of ecclesiastical dumb-bells in the form of a cross; and from the cap an absurd gilt weathercock pecks at the sky.

I thought of the noble ruins I had seen in various parts of the country, the shorn Sampsons, defiant, tragic . . .

But this—this hermaphrodite . . .

O the bitter shame of it!

THE MILL, MONTREUIL-SUR-MER

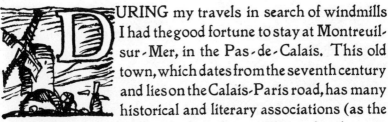
DURING my travels in search of windmills I had the good fortune to stay at Montreuil-sur-Mer, in the Pas-de-Calais. This old town, which dates from the seventh century and lies on the Calais-Paris road, has many historical and literary associations (as the guide books say) that do not directly concern me for the purpose of this book. Windmills, and the people who live in the shadows of windmills, are what I have sought throughout; the nicer, more precise accounts of places where the windmills are I leave to more erudite and, possibly, sincerer persons than myself. Let my course be that of a thistledown, blown willy-nilly, finding a joy here and there in unpurposed movement!

My bedroom in the Hotel de France was that supposed to have been occupied by Sterne and mentioned by him in *The Sentimental Journey*. The *patron* told me this with swift gestures and anticipatory glances. My illustrious countryman, *n'est-ce pas?* I felt rather a worm during the rest of my stay there, and, to make matters worse, the fellow was never tired of talking about it... *Votre grand compatriot* this, *votre grand compatriot* that... till I grew to hate the name of the amorous ecclesiastic. I am an honest pedlar in windmills, while he—he wasn't even respectable! The *patron* knew it, I imagine; he spoke in a manner that seemed to hint at possible disclosures concerning a sticky relative; he smiled diabolically. For the first time in my life I enjoyed Thackeray's priggish condemnation of the parson, and felt, as he must have felt when writing it, magnificently virtuous.

My bedroom overlooked the courtyard. It was a fine room, and the walls were covered with an old Greek paper; I spent many thrilling moments walking to and fro, speculating on the possible origin of the paper, for I could learn nothing definite from the *patron*. Neither could I learn anything about windmills, although there were many watermills in the district. He advised me to pay a visit to Jean M——, the baker, who enjoyed the reputation of being a compendium of useless knowledge. He was so crushingly big and so enormously fat that the large curved pipe which dangled from his mouth looked like a mere acorn. He dwarfed everything in the vicinity. Behind his bulk the fairly large shop looked for all the world like a child's toy, and his wife, by no means *petite*, might indeed have been formed from one of his ribs. I approached the gigantic organism gingerly, fearful of what vocal possibilities might be hidden within. His answered "Bon soir," however, was uttered in the dulcet voice of a eunuch.

"*Comment, Monsieur*, windmills? . . ."

And after a moment's thought:

"No; there are no windmills round here. Watermills, yes, many . . ."

The baker was a scholar and a student of Rabelais, and he asked me if I remembered the Mill of Myrebalais, mentioned in the first Book of Gargantua. The amiable Pantagruelist rambled off into stories that warmed my blood and brought peals of candid laughter from his wife.

"Ah, yes, there is a windmill off the Paris road—a malodorous place; but you may find it interesting."

I listened for a time to more stories, all touched with the spirit of the immortal François—rollicking, full-blooded anecdotes, full of sap and fire; and after a boisterous salutation, I bid farewell to the mountainous Jean and went off in search of the Mill.

It stood, as he told me, by the Paris road, and in a state of utter decrepitude, ravished by whoring winds and pregnant with fungi. But it is not entirely deserted by humans, for there are those who are glad to creep into its old body for shelter from the bitter weather and the dewy nights. The tramps and roadsters, the outcast and destitute have turned the Mill into a lodging-house; and those who seek sensations from the *macabre* might enter the place on a dark night and hear the groans and the snores, drink the foul air, and view the pallid, tortured faces and twisted forms of the sleepers under the light of a dip. It is one of the bare, black bones of our civilization. And this old Mill, which aforetime ground grain to make plump the fair flesh of earth, now, in extreme old age and impotency, provides what scant shelter it can for the rotting, gangrenous bodies of the oppressed.

The Mill stands on a fine point of vantage, and it is probable that the great Vauban used it as a post of observation when building the walls of Montreuil. Like most of the finer examples of French windmills, it is a good height—sixty feet or thereabouts—and covered with small tiles which give a queer decorative effect, and from a distance it looks like a monstrous rotting fungus covered with fish-scales.

There is nothing in the world more pitiable than this forlorn and befouled Mill, nothing more tragic or indecently degraded; shunned by the country folk like a plague, it is the home of dark, nocturnal creatures, scarcely human, all desire and instinct to find some place to hide from their fellow-men and to seek shelter from a world which chills and kills.

THE MILL, COLOGNE

COLOGNE

N turning over some old notes on Italy, it struck me that in all the journeys I have made in that country I do not remember having seen a windmill. It is a most remarkable fact. Of course, there may be hundreds of windmills, dotted here and there, among those innumerable cliff-clinging hamlets that look like groups of cattle grazing in the blue pastures of the sky. But I did not remember a single windmill anywhere in the country. Watermills, yes, especially in Piedmont, Lombardy and Venezia—things of beauty, as motor-cars are beautiful, cold, precise, inhumanly perfect, but without charm for me. I began to conjure from memory all the out-of-the-way places I have visited, thinking the while: There must be a windmill somewhere in Italy. But, no, I could not call one to mind. . . And I grew very angry; for I felt sure that there must be some windmills in a land that never failed to furnish me with anything I desired, be it sun, wine, mud, blue, or the odours of men and the earth. . .

I threw my notes aside and fell to thinking of windmills in Spain, in France, and in Germany. And then my mind stopped. Germany, Cologne. . . Cologne. The Mill of God. . .

If anyone should say, "The Mills of God grind slowly—," I am instantly transported to Cologne, and I see the Mill of God on a height overlooking the Rhine. That is what I have always called it; and I hope it isn't blasphemous.

It is very strange how one's mind in passing through

intellectual and religious experiences will seize the nearest symbol, however trite it be, and hold to it even after such experiences have ceased consciously to be a part of one's life. For instance, I never think of God without thinking of a lamplighter. I mean the old-fashioned village lamplighter who carried a ladder and a box of matches; and as he used to buy his own matches the village was usually in darkness on windy nights. There was such a one in the village where I was born. His name was Withers. He was a tall man with hollow temples and sombre eyes and an enormous white beard. As a boy the story of Jacob's ladder stirred my imagination profoundly. I used to lie in bed at nights pondering this wonderful golden vision. I could see the shining ladders and the tripping angels; I could hear the rustling of their wings and the strains of heavenly music (*White Wings they never grow weary*, by the way). And high above the ladders, beating time with his arms, was old Withers, wearing a nautical cap and white jacket, out of the pocket of which hung a chamois washleather.

Time has dealt that symbol a hard knock. On walking down the village street, years later, with a somewhat fanatical aunt, we passed by old Withers who was carrying his ladder and wearing his blue cap and white jacket.

"Ah, Withers," I murmured; "dear old chap."

"What is that you say?" demanded my aunt. "Dear old chap indeed! He is a disgrace to the village. He is an atheist."

But no aunt can desecrate the Mill of God. It is too high and passionless and remote from human affairs. It is built on one of the towers of an old Gothic church, and I know that its stones are stained with the blood of the damned. They ought to be, anyway. I haven't examined them because it is almost certain that they are cleaned after grinding the children of Beelzebub. It is the sort of thing that pious Germans would do: they never did like a mess.

The Mill of God is unique. How it came to be erected there I cannot conjecture. I only know that the audacious beauty of it takes one's breath away. Indeed I doubt whether there is a more beautiful windmill in Europe; and the strange thing about it is that practical Hans has created it. I suppose it was the fine wind position that induced him to build a mill on the bones of an old fane... Yes, just that... the elevation.

It stands there above the smoothly-flowing Rhine, a menace to wrong-doers. Never mind about the corn it grinds. Let us think of the souls of men. It is a symbol of retribution, high, remote, inscrutable.

Perhaps Hans used it as a searchlight during the war. I don't know. If he did it must have been terrible to behold, black against a stormy sky, the cold white fire of its wrath stabbing the gloom.

It must have been hungry, too, in those days. The children of Beelzebub were closely locked in council chambers and palaces; they wore medals on their breasts and had armed guards to protect them; and the blood of youth was being spilt in order that they might escape the Mill of God.

Those times are no more. But the children of Beelzebub are eternal; and the old Mill has much work to do.

"The world rolls on for ever like a mill,
   It grinds out death and life, and good and ill..."
O Mill of God, grind on with the grinding world!